Investigating Waves

Reader

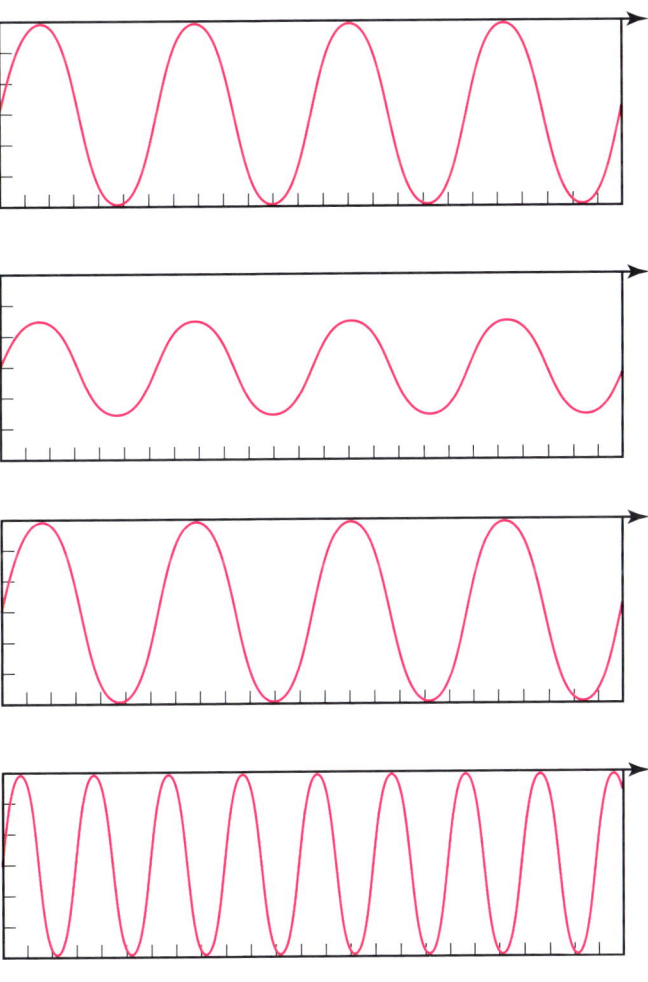

Copyright © 2019 Core Knowledge Foundation
www.coreknowledge.org

All Rights Reserved.

Core Knowledge®, Core Knowledge Curriculum Series™,
Core Knowledge Science™, and CKSci™ are trademarks
of the Core Knowledge Foundation.

Trademarks and trade names are shown in this book
strictly for illustrative and educational purposes and are
the property of their respective owners. References herein
should not be regarded as affecting the validity of said
trademarks and trade names.

Printed in Canada

ISBN: 978-1-68380-522-9

Investigating Waves

Table of Contents

Chapter 1	**Describing Water Waves**	1
Chapter 2	**Sound, Energy, and Change**	5
Chapter 3	**Light Waves**	13
Chapter 4	**Invisible Energy**	23
Chapter 5	**Codes and Signals**	27
Chapter 6	**Using Signals**	33
Glossary		41

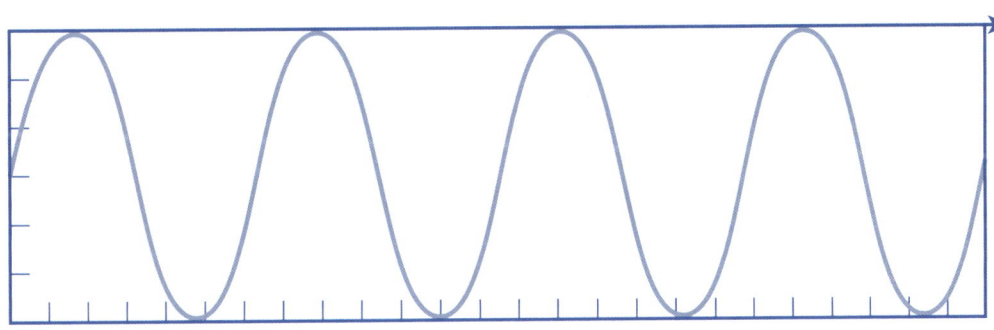

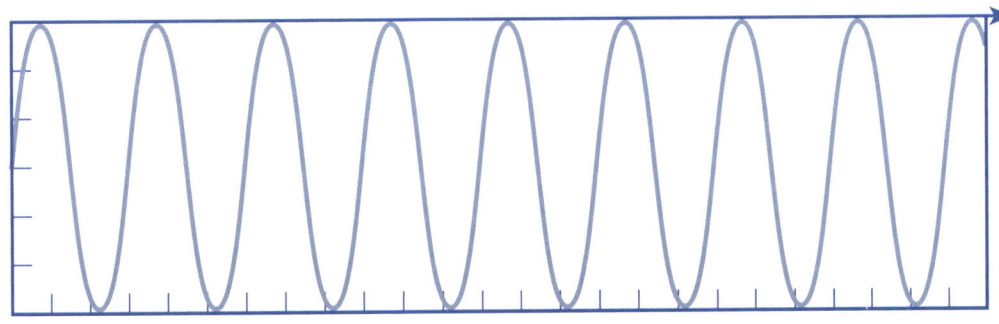

Describing Water Waves

Chapter 1

Look at the surface of the puddle of water. In the picture on the left, it is as smooth as glass. There is no motion in the water. No visible changes are happening on its surface. In the picture on the right, moving feet introduce energy of motion.

Big Question
How does the energy of water waves cause a change?

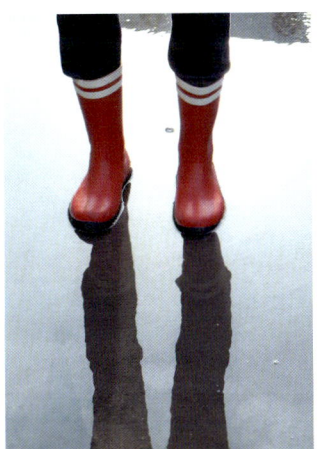

The boots contact the water, and energy is transferred from the boots to the water and causes a change. What change does it cause? It causes a disturbance in the water that we can see. A **wave** forms that moves out through the water in all directions. Water waves move away from the place they start and transfer energy from one place to another in predictable patterns.

Vocabulary

wave, n. a disturbance that transfers energy through matter or through space

Word to Know

A *disturbance* is an interruption of stillness.

Energy from Wind Produces Waves in the Ocean

Imagine that you are in the middle of the ocean. What would you expect to see? Water in the ocean is never still. The surface of the water is always rippled with waves. Some of the waves are low and gentle. Other waves become very tall and steep.

What causes ocean waves? The waves you can see on the ocean's surface are caused mostly by wind. Wind is moving air—it has energy of motion. When the air moves across the surface of the water, the wind pushes on the ocean, and energy transfers to the water. As the air continues to move, more and more energy is transferred to the water, sometimes creating giant waves! This motion energy then moves through the water as waves.

As waves move through water, objects floating on the water's surface bob up and down. A surfer rises as the wave pushes the board up. The surfer moves forward as the wave pushes the board toward the shore.

Energy transfers through the water as the surfer floats on the surface. The movement of the water causes the position of the surfer to change.

Waves Have Characteristics

A wave is not a random transfer of energy from place to place. Waves display regular patterns. If you look carefully at a set of waves, you can see that they have a repeating structure. You can identify different wave parts to describe a wave by its properties. Identifying and measuring different characteristics of a wave is especially helpful when scientists compare different waves.

> **Vocabulary**
>
> **crest, n.** the highest part of a wave
>
> **trough** /trof/, **n.** the lowest part of a wave
>
> **wavelength, n.** the distance from one crest to the next crest of a wave
>
> **wave height, n.** the vertical distance from the top of the crest to the bottom of the trough of a wave

Let's look at a typical water wave from the side. Like most ocean waves, this wave was started when wind pushed on the surface of the ocean. The highest part of a wave is called the **crest**. The lowest part of a wave is called the **trough**. The length of a wave, from one crest to the next crest, is called a **wavelength**. The distance from the top of the crest to the bottom of the trough is called the **wave height**.

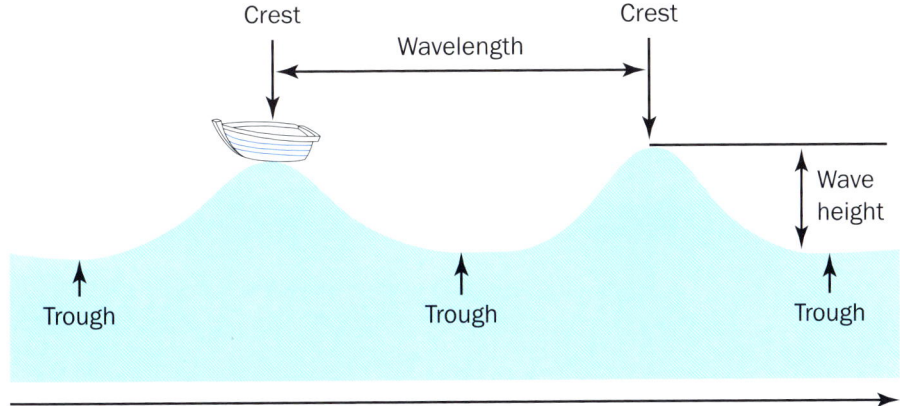

In this wave, energy is moving through the water from left to right, from in front of the boat to behind it.

Waves Differ in Size and Speed

Different size waves have different properties. Crests can be high or low. Troughs can be deep or shallow. Wavelengths can be long, with a large distance from one crest to the next. Or wavelengths can be short, with crests occurring more closely together.

Scientists and engineers use models of waves to help them discuss different kinds of waves and the different kinds of change that they can cause. Big water waves cause greater changes than smaller waves. A wave with a high crest, a deep trough, and short wavelength can toss a boat about for an unpleasant ride!

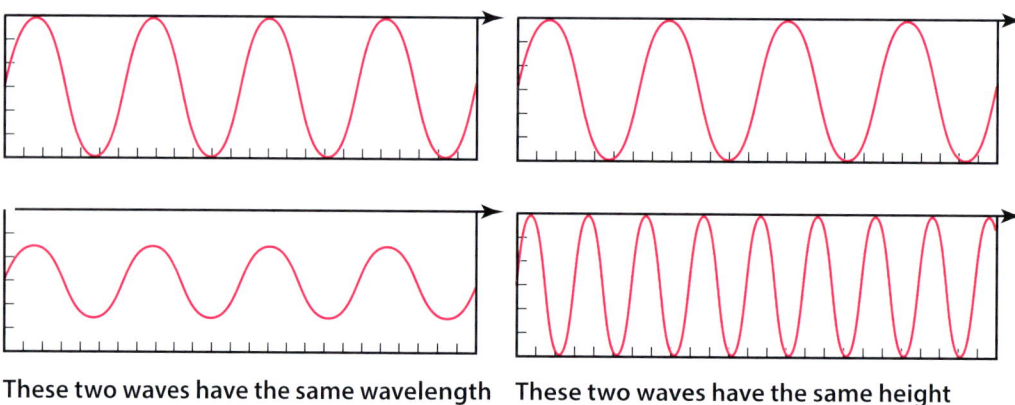

These two waves have the same wavelength but different heights.

These two waves have the same height but different wavelengths.

The speed of energy transfer through water varies as well. Several crests may pass by a given location in a short time, meaning the wave is moving quickly. Or fewer crests may pass by a given location in the same amount of time, meaning the wave is moving slowly. Look at the bottom wave pictured in each pair of images on this page. In which of these waves do the crests occur more frequently? How do you know?

Sound, Energy, and Change

Chapter 2

Place your fingertips gently against your throat. Now say the word **vibrate**. You can feel your vocal cords vibrate in your neck. When something vibrates, it moves back and forth quickly. Even if the **vibrations** cannot be seen, we can often hear evidence that something is vibrating—sound!

When a vocal cord vibrates, it transfers its motion energy to the surrounding air. Then, the surrounding air vibrates, too. The tiny, invisible particles that make up the air bump into each other. They transfer motion energy from air particle to air particle in a pattern. The energy of this vibrating pattern of motion moves through the air as a **sound wave**.

Big Question

What is the relationship among vibration, sound, and energy?

Vocabulary

vibrate, v. to move back and forth quickly

vibration, n. the motion of an object or material that is vibrating

sound wave, n. a transfer of energy through a material as it is disturbed by vibrations

You cannot see air particles, but sound energy makes them vibrate and bump into each other.

Sound Waves Must Move Through Matter

Air is not empty space. Air is a type of matter, a gas. Like all gases, air is made up of tiny particles that are too small to be seen. Because air is matter, a vibrating object can transfer its motion energy through air as sound waves. Imagine a large bell ringing far in the distance. A mallet or hammer may hit the bell, causing it to vibrate. The matter in the air around the bell vibrates, and you hear the bell!

Sound waves can transfer energy through liquid and solid matter, as well as through gas. In fact, sound travels very well through water. For example, whales make certain sounds in the ocean that other whales can detect hundreds (maybe thousands) of miles away!

Sound also travels easily through solid matter. Tap a pencil gently on your desk, and listen to the sound that reaches your ear through the air. Then, place your ear against the surface of the desk, and tap again. The sound seems bigger and louder when you hear it through the solid desk.

Word to Know

Matter that sound energy passes through is called a *medium*. Solids, liquids, and gases are each a type of medium.

Volume and Pitch Are Characteristics of Sound

Some sounds are very strong and loud. Other sounds are faint and quiet. Some sounds, such as the song of a whale, travel far and can be heard over a very long distance. Other sounds, such as a whisper, barely travel across a small room. The loudness of a sound is called **volume**. Loudness is also called **intensity** and is measured in decibels (dB).

The quality of whether a sound is high or low is called **pitch**. We use words such as *booming* and *deep* to describe some pitches. And we describe other pitches with words such as *squeaky* and *sharp*.

The sound quality of pitch is directly related to the **frequency** of a sound wave. Frequency is the number of times a sound wave crests over a period of time. The greater the frequency, the higher the pitch.

Vocabulary

volume, n. the way humans perceive loudness from the intensity of a sound wave

intensity, n. the measurable strength or power of a vibration

pitch, n. the quality of sound that is described as high or low and is related to a wave's frequency

frequency, n. the number of times a wave peaks over a period of time

Word to Know

Vary means to differ. When something varies, that means it is different from one example to another. The loudness and pitch of your voice can vary depending on how you vibrate your vocal cords.

Diagrams Model Characteristics of Waves

A diagram of a sound wave's characteristics can suggest whether a sound is loud or soft, or deep or sharp. Scientists call this type of model a wave line, and it takes the form of a graph. Let's look at wave lines of two different sound waves, one that represents a loud sound and another that shows a soft sound.

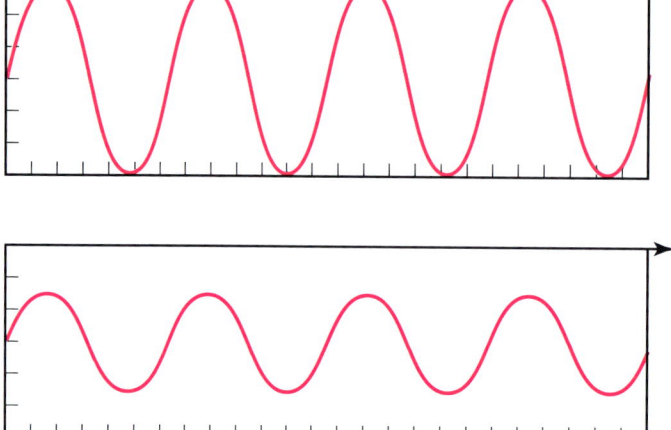

The wave with greater height (the biggest difference between the crest and trough) models a louder sound. The wave line with the lower crest represents a quieter sound. Of these two wave lines, which represents a louder sound?

Next, let's look at a wave line of two other sound waves, one that represents a high-pitched sound and one that represents a low-pitched sound.

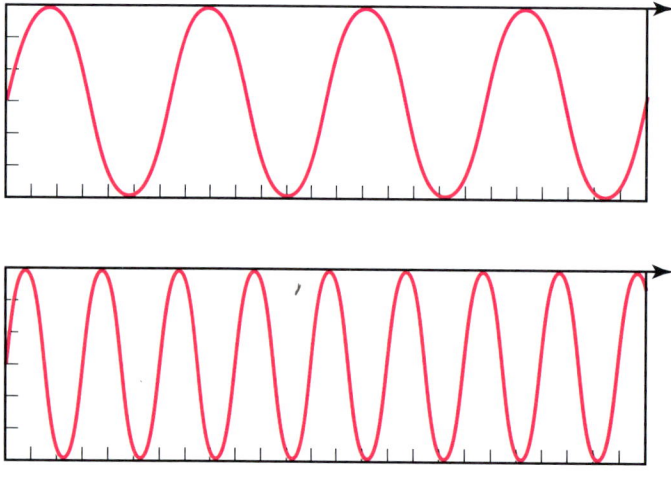

The wave line with less frequency represents a sound with a lower pitch. The wave line with the greater frequency is a sound with a higher pitch.

Sound Waves Travel at Measurable Speeds

When someone speaks to you across a room, the sound seems to reach your ears almost immediately as the words are spoken. But it does take time for the vibrations to move through the air from one place to another. Sound moves through air at a speed of about 1,124 feet (343 meters) per second. That means a sound wave can travel over three football fields in one second when the conditions are right.

Air temperature affects the speed at which sound can travel through air. The speed of sound is a little faster in hotter air and a little slower in colder air. Sound waves also move more quickly through liquid matter than they do through gas. And sound waves move even more quickly through solid matter than they do through liquids.

The Brooklyn Bridge is about a mile long from one end to the other. A car travelling thirty miles per hour could cross that distance in two minutes. It takes a sound about five seconds to travel across the length of this bridge.

Think About Distance and Speed

Think about how fast you move in a car on a highway. If you drive sixty miles per hour (mph), that is one mile per minute. A mile is equal to 5,280 feet.

Sound can travel one mile in five seconds! In other words, sound travels about twelve times faster than you do in a car on a highway.

Sound Energy Can Cause Changes

Remember that sound is a form of energy, and energy causes change. Have you ever been in a room when an airplane passed over, flying low enough to rattle the windows? Or during a storm, you might have heard thunder that was such a loud *BOOM* that you could feel it as well as hear it.

Sound at a particular pitch can shatter certain glasses. A sound that can break a glass does not have to be extremely loud. The sound's pitch, when matched just right to the glass, causes the glass to vibrate in a way that can make it break apart.

Sound and Medicine

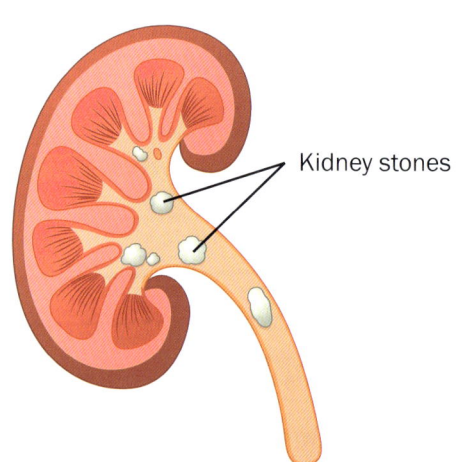
Kidney stones

People can develop a medical condition known as kidney stones. This is when your kidneys hold calcium that turns into pebble-sized deposits. It can be painful, and it used to require surgery to remove the stones. A way has been developed that uses sound waves instead of surgery. The sound waves go from the source into the body and break the kidney stones into smaller pieces.

Word to Know

An object that vibrates and creates a sound is called the *source* of the sound. Speakers, tuning forks, and instruments all have parts that vibrate to be the source of sounds.

Animals Detect and Use Sound

Many living things detect sound energy. For what we call the sense of hearing, in most cases, hairs inside or outside the animal sense the sound vibration.

In humans, sound is transferred from air to liquid before we hear. The human ear funnels sound to the ear canal. The sound moves the eardrum, which pushes fluid inside the inner ear. This makes the hairs inside the cochlea vibrate, and we hear sound.

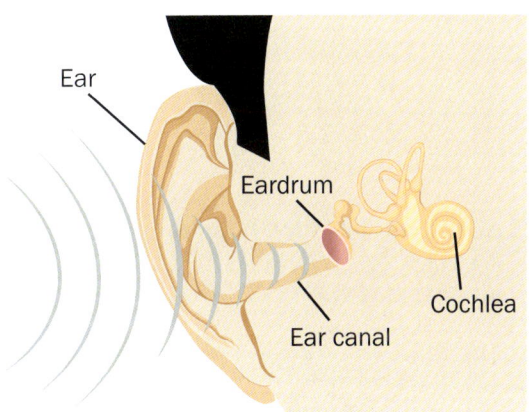

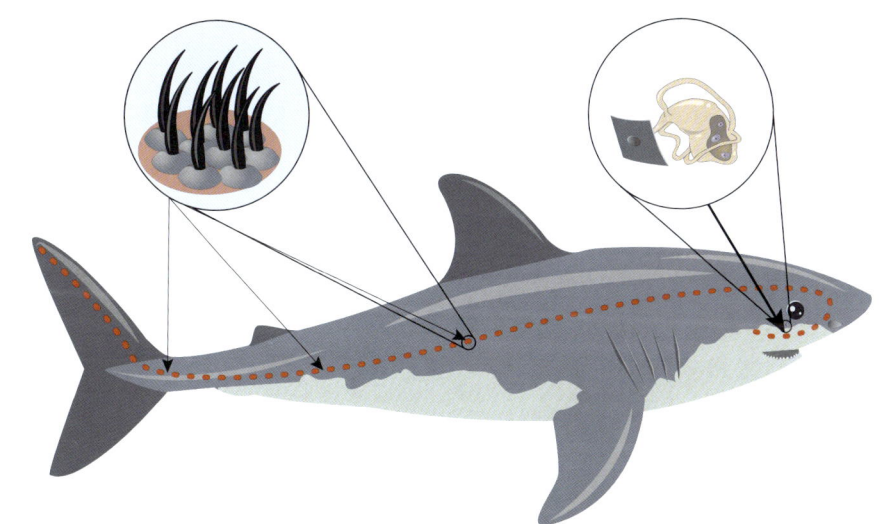

A shark has two ways to hear. It has a hole just below its eye that funnels sound to a structure inside its skull. Inside there are hairs that sense vibration. A shark also has small pores along the length of its body that have hairs inside. These hairs also sense vibration.

A spider doesn't have ears, but it can hear just fine. Much like the pores on a shark's body, a spider has hairs on the outside of its body that vibrate to alert the spider about prey and predators.

Sensitivity to Sound Varies

Dogs can hear sounds at pitches too high for people to hear. Elephants can hear sounds at pitches way too low for people to hear. Some animals detect sound in different ways, with very different structures from eardrums. For example, many insects have tiny hairlike structures on their antennae, legs, or wings that enable them to feel vibrations in the air.

Animals that live in water can detect sound as well. Many aquatic animals can detect sound over much longer distances than most animals that live on land. Because sound waves travel at different speeds through water and air, these animals have structures that help them hear different pitches and intensities of sound that humans may not be able to hear or understand.

The ability to detect sound is very important for many animals' survival. The sense of sound may warn animals about the threat of danger. Many kinds of animals that are nocturnal, or are most active during the night, have highly sensitive senses of hearing. Many animals produce sounds, too. Animals can make sounds to warn others of a threat or to attract a mate.

The greater wax moth has the greatest sound sensitivity of any animal—better than dogs or owls or bats that prey on them. A dog's hearing is about four times better than a human's, but a moth's sense of hearing is about fifteen times better.

Light Waves

Chapter

3

What do you notice about the forest floor in this picture? What is the difference between the bright green areas and the parts that appear to be darker green? The brighter parts are areas that are well lit by sunlight. The darker parts provide evidence that some of the **light** from the sun is blocked by the trees.

Big Question

How does light behave?

Vocabulary

light, n. a form of energy that can transfer through empty space and can make things visible

Light is a form of energy. Light provides evidence that energy is transferred from one place to another. Like sound, light energy moves in patterns that we call waves. However, unlike sound, light does not have to transfer through a medium. Light can transfer through empty space that contains no matter at all.

Light Waves Come from a Source

Stars, such as the sun, and objects such as light bulbs and glow sticks are examples of things that emit their own light. **Light sources** are objects that convert some other kind of energy to light energy. The sun gives off light because of the reactions that occur within the star. Light bulbs transform electrical energy to light using a filament that glows brightly when electricity passes through it. Glow sticks combine chemicals to give off light.

> **Vocabulary**
>
> **light source, n.** an object that gives off its own light

Fire is another light source. When a material burns, a chemical reaction occurs that gives off light and heat. For many thousands of years, people used only fire to shine light in dark spaces. Eventually, engineers designed ways to convert electrical energy to light, using reliable light bulbs.

Sunlight travels about 93 million miles (150 million kilometers) through empty space as it transfers energy from the sun to Earth.

Light Travels Quickly

Light energy travels in patterns, called **light waves**. Light waves transfer outward in all directions from a light source. Light waves are also known as *radiation*. Light waves radiate (travel in waves) at a speed of about 186,000 miles per second, or about 300,000 kilometers per second. That's remarkably fast!

> **Vocabulary**
>
> **light wave, n.** an energy disturbance that transfers, or radiates, light

Think About the Speed of Light

Light could travel around Earth at the equator 7.5 times in one second. (However, light waves travel in a straight line, so you would need a lot of mirrors to bounce that light around Earth!)

Think about the math. How much time does it take for the sun's light to reach Earth?

Earth's average distance from the sun is about 150 million kilometers.

$$\frac{300{,}000 \text{ km}}{1 \text{ second}} = \frac{150{,}000{,}000 \text{ km}}{? \text{ seconds}}$$

(*The answer is 500 seconds.*)

$$\frac{60 \text{ seconds}}{1 \text{ minute}} = \frac{500 \text{ seconds}}{8 \text{ min } 20 \text{ sec}}$$

How does the speed of light compare to the speed of sound? Think back to the example about sound taking about five seconds to travel across a one-mile-long bridge. In that same five seconds, light could travel thirty-seven times the distance of Earth's equator!

Light Interacts with Matter

Light sources emit their own light. Other objects may appear bright to us, but they do not give off light themselves. These objects **reflect** light from other sources. For example, the moon does not emit its own light—it reflects light from the sun. Sunlight bounces off the surface of the moon, making the moon appear bright to us against a darker sky.

Light interacts with matter in different ways, depending on the properties of the object. Light energy can pass through some objects, but it reflects off others, such as the moon. Light passes through clear glass, but it cannot

The water and the vase are both transparent. The plant is opaque.

pass through a wooden table. The wooden table casts a shadow behind it, similar to the trees in the forest on page 13. Materials such as glass that allow light to fully pass through them are called *transparent*. Air is a transparent gas. You can see right through it. Water is a transparent liquid. Light passes right through clear water. And clear glass is a transparent solid. Scientists and engineers use **transparency** as a characteristic to describe how matter interacts with light.

> **Vocabulary**
>
> **reflect, v.** to bounce off of
>
> **transparency, n.** a property of matter that allows light to pass through it

Some materials only allow some light to pass through them, but not enough that we can see clearly what is on the other side. These materials are called *translucent*. Translucent materials block some light and produce slight shadows.

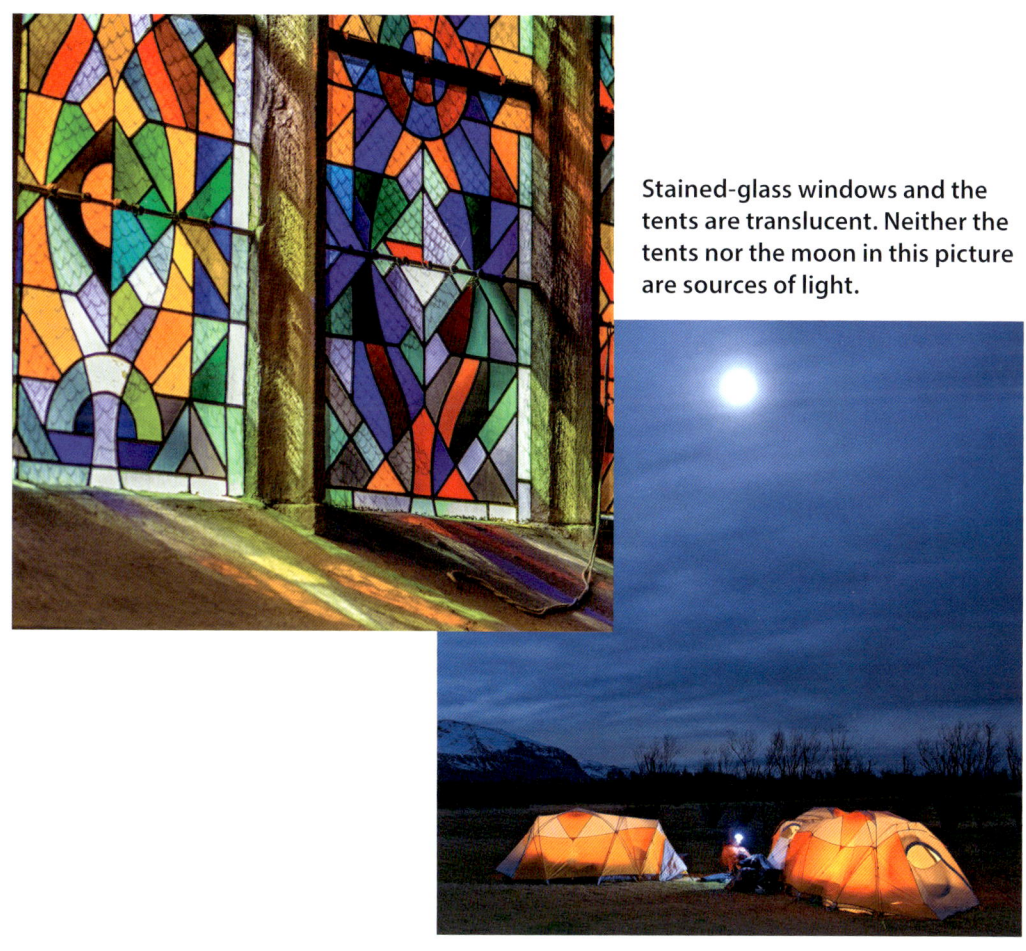

Stained-glass windows and the tents are translucent. Neither the tents nor the moon in this picture are sources of light.

Still other materials allow no light to pass through them at all. These materials are called *opaque*. Opaque materials, such as wooden tables, block all light and cast sharper shadows than translucent materials. Understanding the property of transparency can help scientists develop solutions to problems such as keeping sunlight out of a building while allowing people to see outside.

Properties of Light Waves Determine Color

What is the difference between a blue shirt and a yellow one? Light waves occur with different frequencies. Different colors are a result of the different frequencies of light. Materials can reflect some frequencies of light but **absorb** other frequencies. The frequencies of light that a material absorbs do not reflect.

The **color** of an object depends on the wavelength of the light that is reflected and enters our eyes. So, what about your blue shirt? It reflects blue light waves.

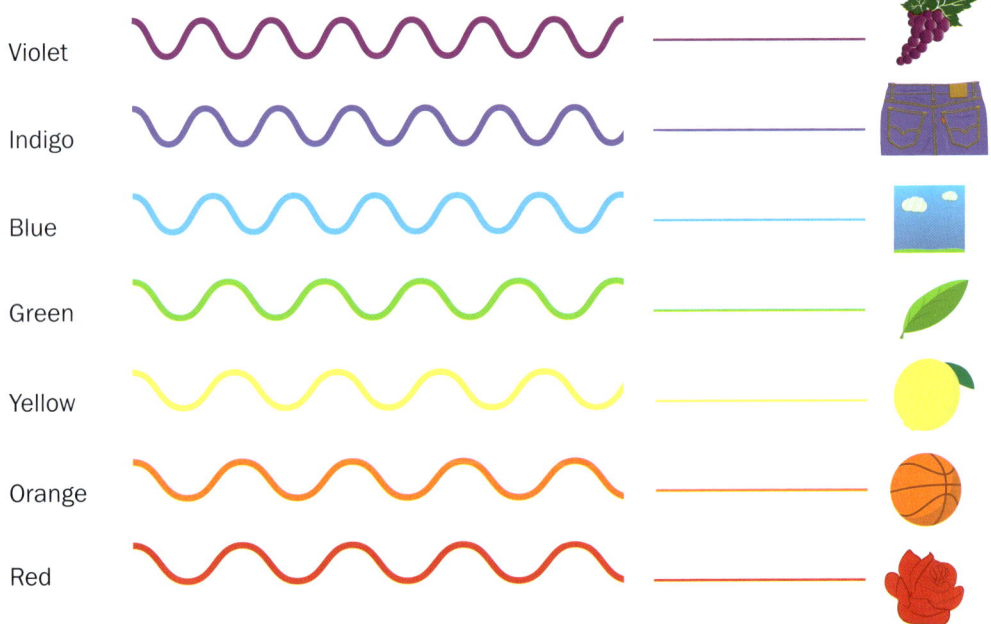

Vocabulary

absorb, v. to take something in and contain it

When light is absorbed, it is not reflected.

color, n. an aspect of light that enables otherwise identical objects to be distinguished from each other through the sense of sight (vision)

For example, one of these squares is blue. The other square is yellow. Color is the only way in which the two squares differ.

Properties of Light Waves Determine Brightness

Now, consider the brightness of different lights that you have seen. The brightness of light is related to the amplitude (height) of its waves. Light waves with greater amplitude have more energy than light waves with less amplitude. Light with higher amplitude is brighter, or more intense. Light with lower amplitude is dimmer, or less intense. This is similar to sound waves. The sound waves with greater amplitude are louder.

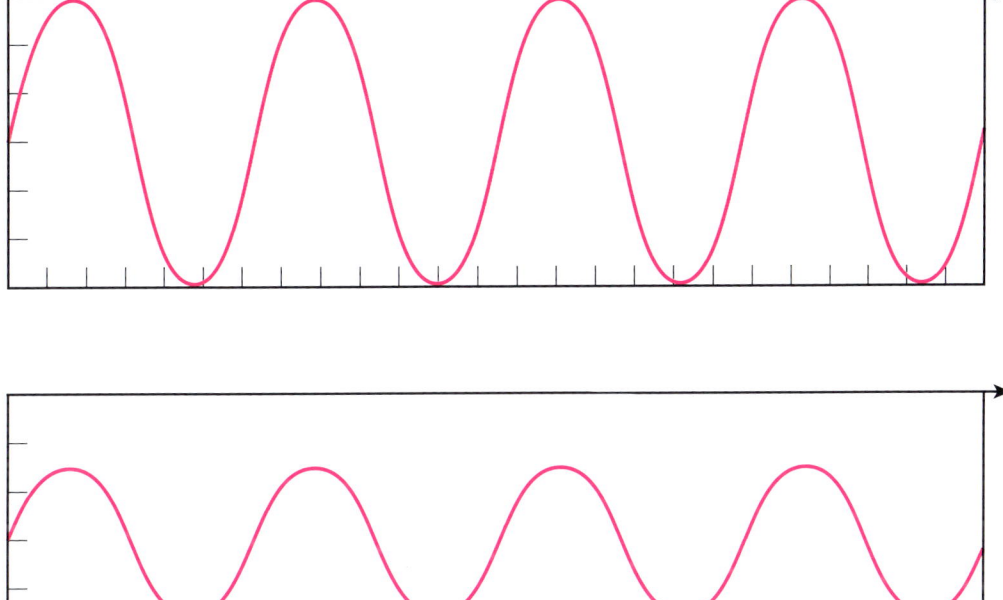

What can you conclude about these two light waves, which have the same wavelength but different heights?

Light Can Be Changed and Cause Change

Light Can Be Changed

Matter can slow light waves down. Passing through different kinds of matter, from one medium to another, slows light waves and causes them to change direction. Light also bends at the corners of barriers. If you know what to look for, you can see evidence of these properties.

Passing through the air and then water slows light down very slightly. This changes where the lemon is seen through the liquid. Predict what happens when you move the glass. The lemon doesn't change. The light reflecting off of it does!

Light passing through a prism at a certain angle separates the wavelengths of light into its different visible colors.

Light Can Cause Change

Remember that light is a form of energy and that energy causes change. Like all forms of energy, light can cause changes in the materials that it contacts. Sunlight on your bare skin can cause a change—ouch!

Light also causes changes in solar panels and in living plants. The changes convert the light waves to different forms of energy. Light transforms to electrical energy in solar panels. It transforms to chemical energy in plants.

Living Things Can Sense Light

Living things detect light waves in various ways. Light makes the sense of vision possible. Eyes are organs that many kinds of animals have that detect light.

How does the human eye work? Light waves enter the eye and cause changes that help the brain sense light waves. Our eyes have special structures that transform light energy to the electrical energy of nerves. These nerves send signals to the brain. The brain decodes those messages and translates them into the images we perceive as sight.

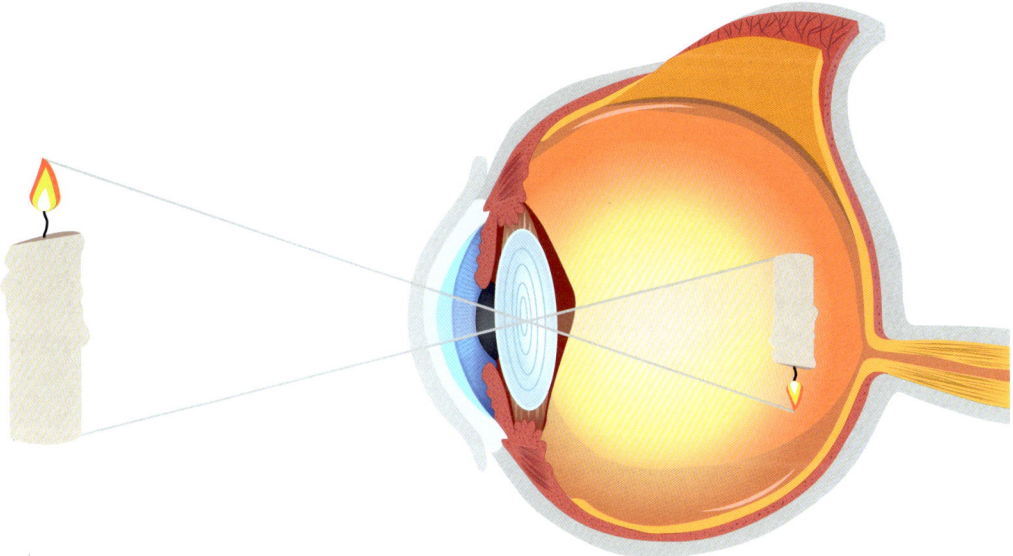

Light waves enter the eye. A nerve in the eye translates the light to an electrical signal that travels to the brain.

Almost all animals have eyes of some type, and many see better than humans. Many insects can sense light waves at wavelengths that humans cannot see. Many birds and mammals can see such dim light with very low intensity that humans would perceive it as total darkness.

This fly has compound eyes. Its eyes are made of many more parts than a human's eyes, and it can see light at more wavelengths.

Owls' eyes, when compared to a human's, are much larger in proportion to their heads. Owls can see much better than people can in low light.

Plants also sense light. They need light to survive. Plants sense the direction of the source of light. Plants can change the direction that their leaves face so they can increase the amount of light they are able to absorb.

All the sunflowers in this field are facing the sun. As the sun moves across the sky, the flower heads will turn to keep following the source of light.

Invisible Energy

Chapter 4

You know that different colors have different frequencies. The color of an object is the reflection of the wavelength for that color. All other colors in the wavelength are absorbed by the object. All the colors and all the wavelengths of light that humans can see are called *visible light*.

When you go outdoors on a sunny day, you need to wear sunscreen to keep your skin from burning. What kind of invisible energy can burn your skin from a distance? Ultraviolet energy waves have a shorter wavelength than visible violet light. What are the other wavelengths that we don't see? Keep reading to learn more.

Big Question

What are some different kinds of light waves?

Violet has the shortest wavelength of visible light.

Violet
Indigo
Blue
Green
Yellow
Orange
Red

Red has the longest wavelength of visible light.

Ultraviolet light waves can cause sunburn.

Infrared and Ultraviolet Light

Some light has wavelengths that are longer than visible red light. This light is called *infrared light*. We can't see infrared light, but special cameras can detect it. Infrared cameras allow us to see warmer and cooler objects in pictures. Infrared light is also used in most television remote controls. A beam of infrared light goes from the remote to the TV, and changes occur when the energy hits the TV. Infrared light is the kind of light that heats matter.

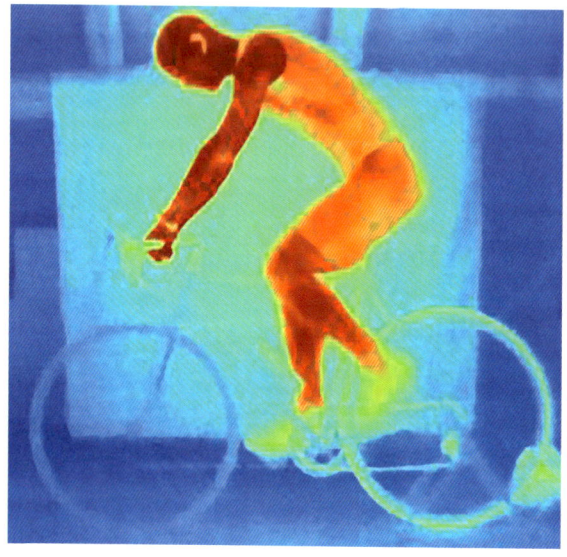

What things in this picture look hottest? Which look coldest?

Some waves of light have wavelengths that are shorter than visible violet light. This light is called *ultraviolet light*. Ultraviolet light from the sun is what causes sunburns. You can get a sunburn from ultraviolet light waves even on cold or cloudy days.

All these wavelengths, including visible light, are examples of light energy, also called *electromagnetic energy*. Keep in mind that all these wavelengths are light energy. Remember that energy causes changes. All of these forms of light are capable of causing changes.

Invisible Energy

Radio waves have the longest wavelengths and the lowest energy. Radio waves can carry signals over long distances, so they are used for communication.

Microwaves have wavelengths longer than infrared light. Microwaves are used in kitchen appliances that quickly heat food.

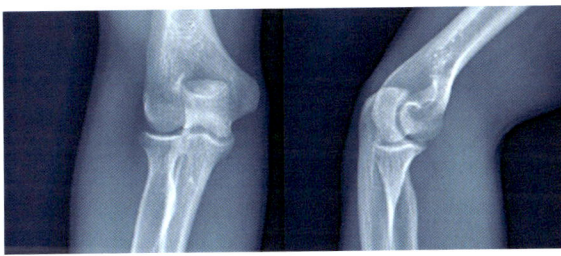

X-rays have wavelengths longer than gamma rays but shorter than ultraviolet light. X-rays can be used to take medical images, such as these pictures of leg bones.

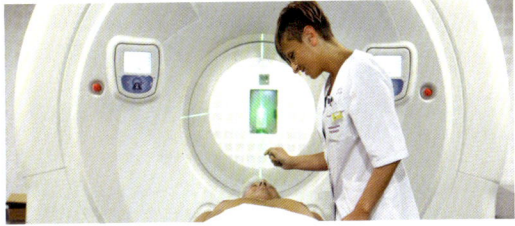

Gamma rays have the shortest wavelength. These rays have a lot of energy. They are used in radiation therapy to help kill certain cells that occur in some types of cancer.

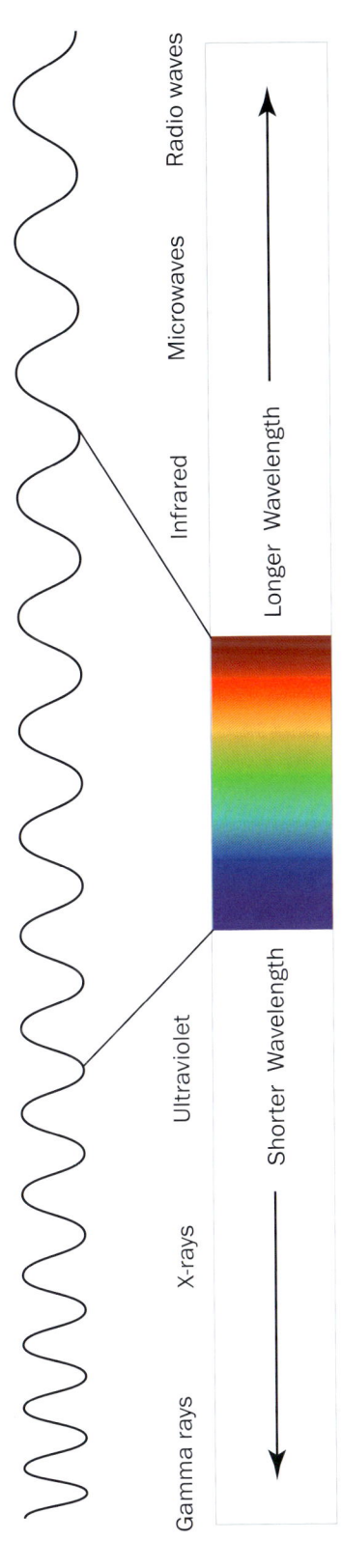

Electromagnetic Radiation Can Be Helpful and Harmful

Many engineering designs use electromagnetic energy to solve problems. Here are a few more handy devices that work using energy waves.

A smartphone with GPS capability is in ongoing contact via electromagnetic waves with a network of satellites orbiting Earth.

Cell phones and wireless internet routers (Wi-Fi) use radio waves. Bluetooth® uses radio waves to connect electronic devices without wires. Two-way radio communications can be very powerful. Radio devices can communicate to and from ships at sea, airplanes in flight, and spacecraft.

Radio waves are also used in radar technology for tracking objects and forecasting weather. Microwaves are used in the network of satellites and devices that make up the Global Positioning System (GPS).

Though electromagnetic energy can be very useful, some forms of radiation can also be harmful. You already know about sunburn. Too much exposure to X-rays can cause some types of cancer.

Certain materials that are called *radioactive* give off radiation. Some radiation may be in the form of gamma rays. Because of their high energy, gamma rays can be deadly to living things. This radiation can damage living tissue. Therefore, radioactive materials must be contained and handled very carefully if they are used for technology.

This hazard symbol is displayed on materials that pose a risk of exposure to harmful radiation.

Codes and Signals

Chapter 5

People communicate with each other in many ways. If you smile at a friend, the friend knows that you are happy. Your smile sends a message that tells how you feel. If you nod, your friend knows that you mean "yes." If you shake your head from side to side, you communicate the message "no." If you wave your hand as you leave, your friend understands that it means "goodbye."

Smiling, nodding, headshaking, and waving are gestures. These are ways that people communicate when they are together. What is the role of light waves in this kind of communication?

Big Question

What is a code?

Word to Know

A *gesture* is a movement of a part of the body to communicate an idea or emotion. Most gestures are made with your hands or your head.

Look at the photo. What emotion or idea is she trying to communicate? How can you tell?

Symbols Are Ways of Communicating

People communicate with gestures. In what other ways do you communicate when you are with others? One way is to speak. You use your voice to make sounds that you and your listener understand as words. The words combine to form messages. The messages contain the ideas you want to share with others. What is the role of sound waves in this kind of communication?

Gestures and speech are ways of communicating, but there are others. You don't even have to be close to communicate. Humans have many ways to send information to others who are far away. The most familiar way you might know to do this is to use written words in a letter, an e-mail, or a text message.

The written word uses **symbols**. The alphabet is a set of letters we use for written languages. Each letter in the alphabet is a symbol. A symbol is a visible object or mark that stands for something else. Symbols represent ideas. Letters are symbols that form words. We can combine letters and words in different ways to record an endless number of ideas.

> **Vocabulary**
>
> **symbol, n.** a visual object or mark that stands for something else

We can then send those symbols to different places or publish them in books or magazines. We can draw them on cave walls or on paper. Because of written symbols, people can communicate over long distances and even over time.

Thousands of years ago, some people used symbols carved into rock to communicate ideas.

Codes Are Ways to Communicate

People use other symbols besides written language to communicate. Many of these are symbols you see frequently. You know what they mean without having to think about them.

A **code** is a **pattern** of symbols. Codes can be patterns that you see, such as letters or numbers, or something that you hear, such as the clicking sounds of Morse code. Morse code uses patterns of dots and dashes that stand for letters and numbers. The dots and dashes are translated into a pattern of long or short clicking sounds. The sounds can be sent through electrical wires. People used Morse code to communicate over long distances before radio communication and telephones were invented.

Vocabulary

code, n. a pattern of symbols that can be used to communicate a message

pattern, n. a regular or repeated way in which something occurs

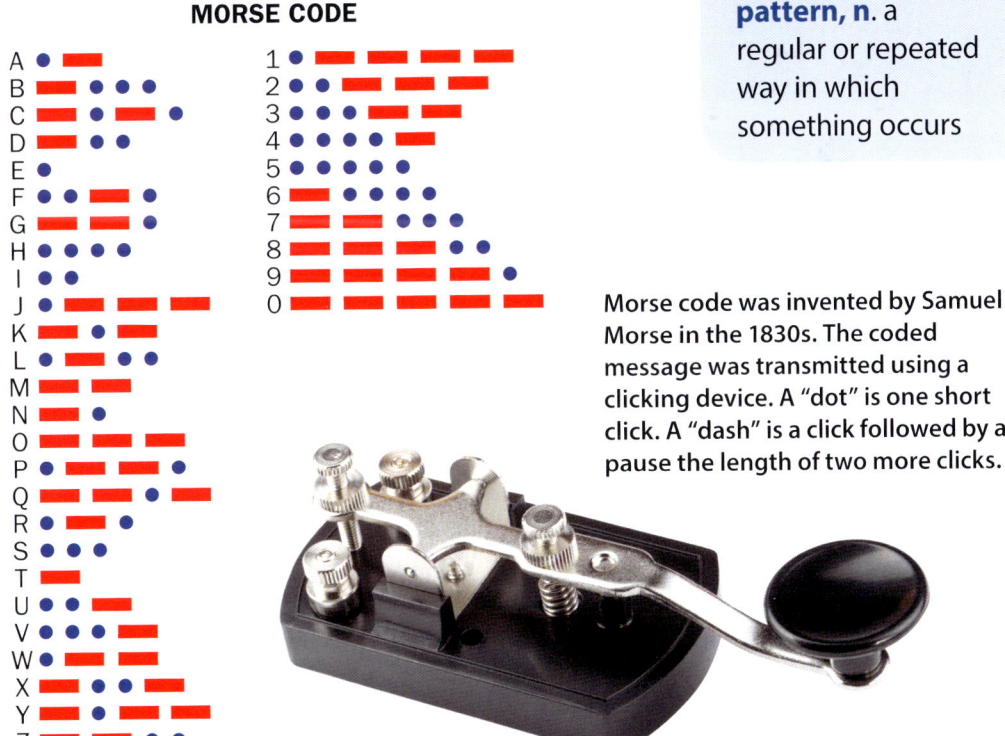

Morse code was invented by Samuel Morse in the 1830s. The coded message was transmitted using a clicking device. A "dot" is one short click. A "dash" is a click followed by a pause the length of two more clicks.

Signals Transfer Information

The clicks that are used in Morse code are sent over long distances. They are one kind of **signal**. A signal is the way that a word, symbol, or code is sent over a distance.

> **Vocabulary**
>
> **signal, n.** a symbol sent over a distance

Signals You Can See

A lighthouse projects a bright, flashing light out to sea. The lighthouse's lamp is the source of the light waves. Sailors detect the light waves and observe any patterns they see. They recognize patterns as symbols or codes. Every lighthouse uses a unique pattern of flashes. Sailors look at a chart to help them decode each signal and find out their exact location.

The light from a lighthouse is a signal, sent as light waves, that keeps boats safe as they approach shore.

Light waves are employed in communicating an idea. A traffic light is another kind of signal. It flashes red, green, or yellow. These colors let drivers know whether they should stop, go, or begin to slow down. Traffic lights help drivers move safely in different directions. They also help keep traffic flowing.

How does the traffic light signal that these cars should stop?

Signals You Can Hear

You learned that each lighthouse has its own unique light pattern. But what about during the day or on foggy nights when sailors can't see the light? Every lighthouse also has a foghorn. A foghorn makes a loud sound that can be heard far from land. Each foghorn makes its own unique pattern of sounds. Sailors can listen to the signal and figure out their location. This is an example of sound waves sent out in a pattern to communicate information.

The shape and size of a foghorn helps its signal be heard far from shore.

How can you use sound to send whole words and sentences? Items such as telephones use electrical energy. When you speak into the phone, your voice makes sounds. These sounds transform to electrical energy that can travel through wires. When the electrical energy reaches the phone on the other end, it changes back to sound.

Think About a Telegraph

You learned about Morse code and how it is a pattern of sounds used to send messages. It is sent by telegraph. A *telegraph* is a device that moves electrical signals from one place to another through wires. Samuel Morse worked with other inventors to develop the telegraph in 1836.

No Wires? No Problem!

When you use a two-way radio, you use it like a telephone, but there are no wires to carry the sound energy from one place to another. So how does it work? When you speak, the two-way radio changes the sound energy from your voice to radio waves. The waves, which are a signal, travel to another two-way radio that is nearby. That radio receives the waves, and the radio waves change back to sound energy. Then your friend can hear your voice.

Many handheld two-way radios usually only work over short distances. However, some can send signals up to 50 miles.

Cell phones work in much the same way. They also change sound energy to radio waves. But with a cell phone, you can talk to someone on the other side of the world. That is because the radio waves bounce from place to place until they reach the person you are talking to.

Cell phones are only one of many ways to communicate. Gestures, symbols, and codes all communicate messages across a distance. The signal transfers a pattern. Often, this pattern is made up of light waves or sound waves.

Using Signals

Chapter 6

People listen to music on a radio when they ride in a car. Or they watch TV when at home. How do these devices work?

Both a car radio and a TV receive a signal that has been broadcast. *To broadcast* means to transmit, or **transfer**, a signal in some form of a wave. Broadcast stations such as radio and television stations send signals in the form of waves to receivers in TVs and radios. The signals transmit **information** such as symbols, pictures, and sounds, and often they do so in the form of waves.

Big Question

How are waves used to send signals?

Vocabulary

transfer, v. to move from one place to another

information, n. types of data such as pictures, sounds, and messages

Broadcasting devices use technology to change sounds or pictures to signals in the form of waves. These wave signals are broadcast in a unique pattern. The pattern is sent as a wave signal from one device to another. That device transforms the wave signal back to sounds or pictures.

33

What Is Analog Information?

There are two ways to send information in a wave signal. You can use analog signals or digital signals.

Analog signals send information using a continuous flow of electrical energy. Patterns of electrical energy travel as a continuous wave whose pattern varies as it travels through the atmosphere or through electrical devices. An old-style phonograph record uses an analog signal when the needle moves up and down on the record to create a flow of electrical energy. That energy powers speakers, and we hear the recording from the record.

> **Vocabulary**
>
> **analog, adj.** describing signals sent as continuous waves

TV broadcasts once used analog signals to send and receive information in the form of radio waves through the atmosphere. Most radio broadcasts still do. These are analog signals because they are a continuous flow of wave energy.

Broadcast towers send the digital and analog signals across large areas.

34

What Is Digital Information?

An analog signal is always a continuous flow of information. The other way that signals can be sent is by using **digital** signals. *Digit* means number. Digital communication turns information into number patterns. Information sent by digital signals is not continuous—it is broken down into digits, or millions of individual bits of information.

> **Vocabulary**
>
> **digital, adj.**
> describing signals sent as separate bits and not continuous

Digital signals exist in discrete bits and often only include the numbers 0 and 1. Information is thus broken into a pattern and is a code. Think of a digital signal as a switch being turned on and off. This code can be sent with or without wires. With a digital signal, information can be sent anywhere. Tablets, computers, and cell phones use digital signals to send and receive information.

In common computers, digital signals are coded in patterns. For example, 01000001 represents the letter *A*. It would take a lot of 0s and 1s to write your name on your paper using this code! Digital signals can be recorded onto CDs, DVDs, and Blu-ray™ discs.

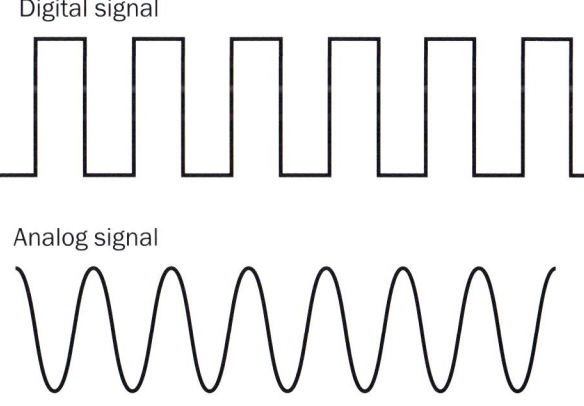

Look closely at the diagrams for digital and analog signals. How are they similar and different?

35

Cell Phones Use Wave Signals to Transfer Information

Imagine using a phone to send a text message. You type the message and press send. Then your friend gets the message. But what is really happening?

Cell phones use digital signals to send messages. A cell phone turns your message into a digital pattern of 0s and 1s.

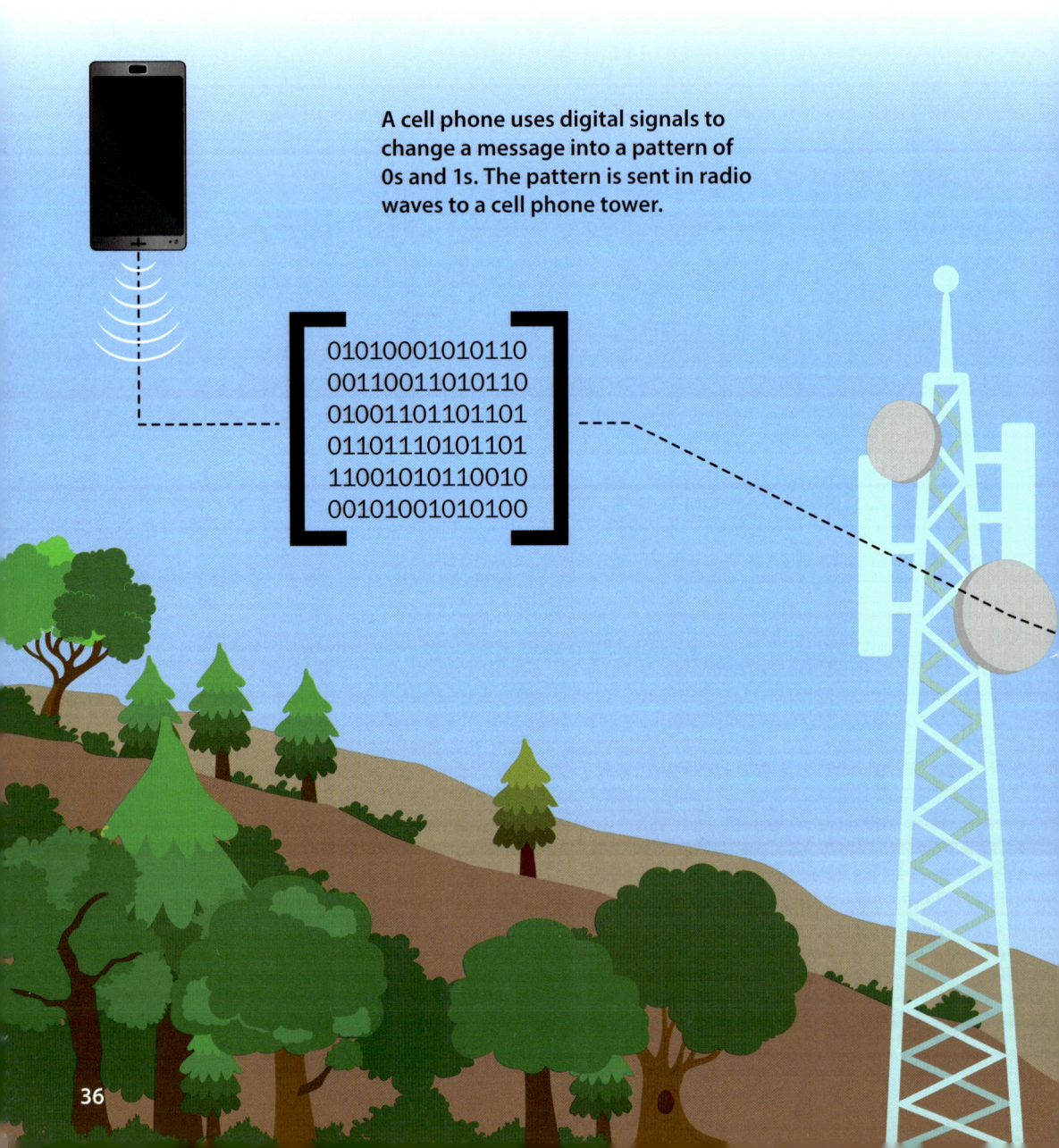

A cell phone uses digital signals to change a message into a pattern of 0s and 1s. The pattern is sent in radio waves to a cell phone tower.

The cell phone then sends this digital signal by radio waves to a cell phone tower. The cell phone tower passes the message on by sending the signal to another tower or directly to your friend's cell phone. Your friend's phone receives the code and changes the pattern of numbers back into a text message that your friend can see.

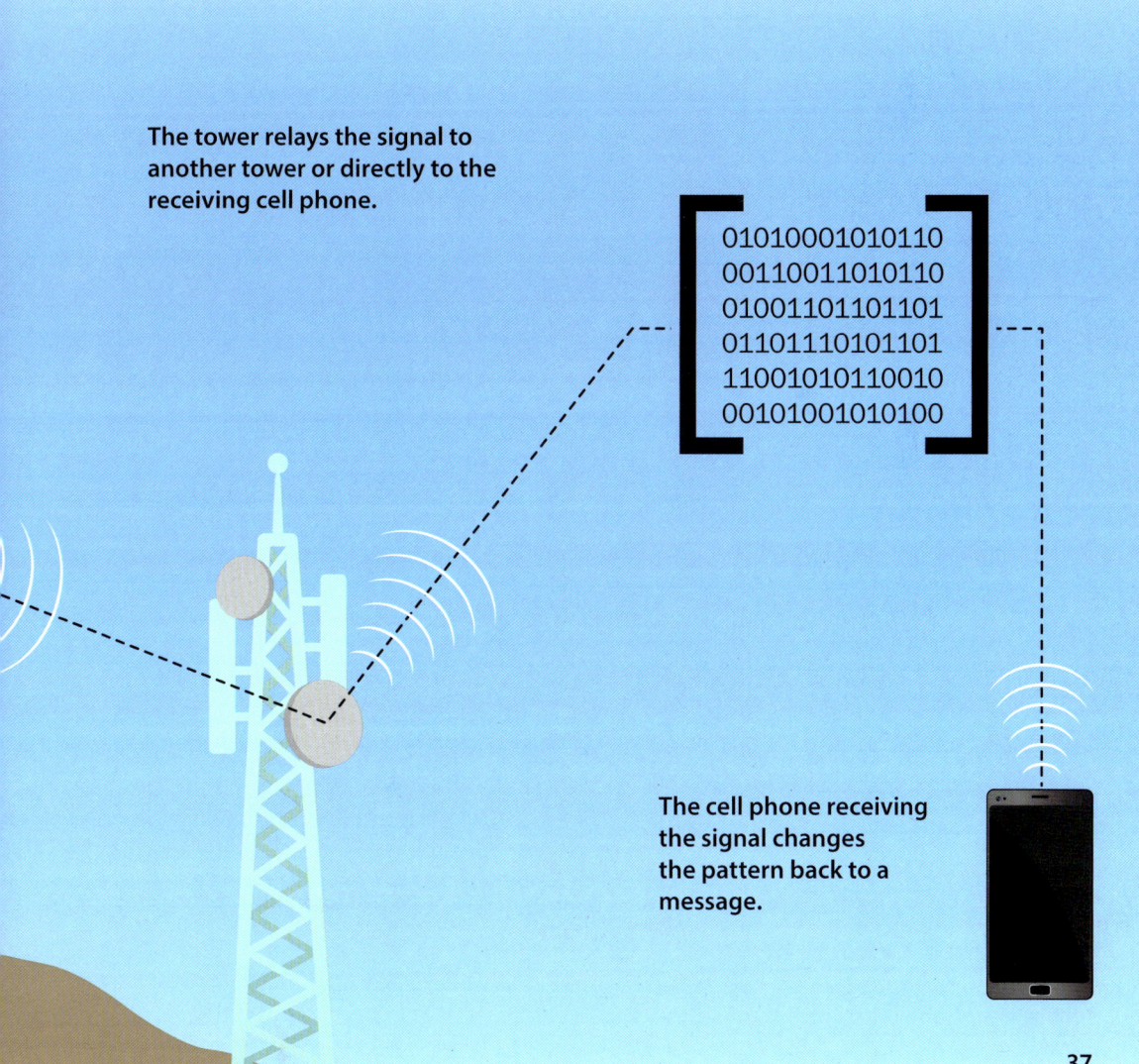

The tower relays the signal to another tower or directly to the receiving cell phone.

The cell phone receiving the signal changes the pattern back to a message.

GPS Uses Wave Signals to Transfer Information

GPS stands for Global Positioning System. You may have used a GPS device or app when traveling or playing a game. GPS can help people find their location or figure out where some other location is. The system began as a tool for people in the military to know exactly where they were when in remote places. It is now used by people everywhere.

Suppose you wanted to know your exact location on Earth. How can the GPS app in your cell phone do that? A device such as a cell phone broadcasts a signal of radio waves to three different satellites.

Each satellite is a different distance from the phone. So it takes radio waves different amounts of time to travel from the phone to each satellite and back. That data can be used to figure out where the phone is located.

The satellites return the radio wave signals back to your cell phone. Your cell phone measures how long it took the waves to travel to and from the three satellites. It then performs calculations to determine your location. Notice that the signal is a simple one, a radio wave.

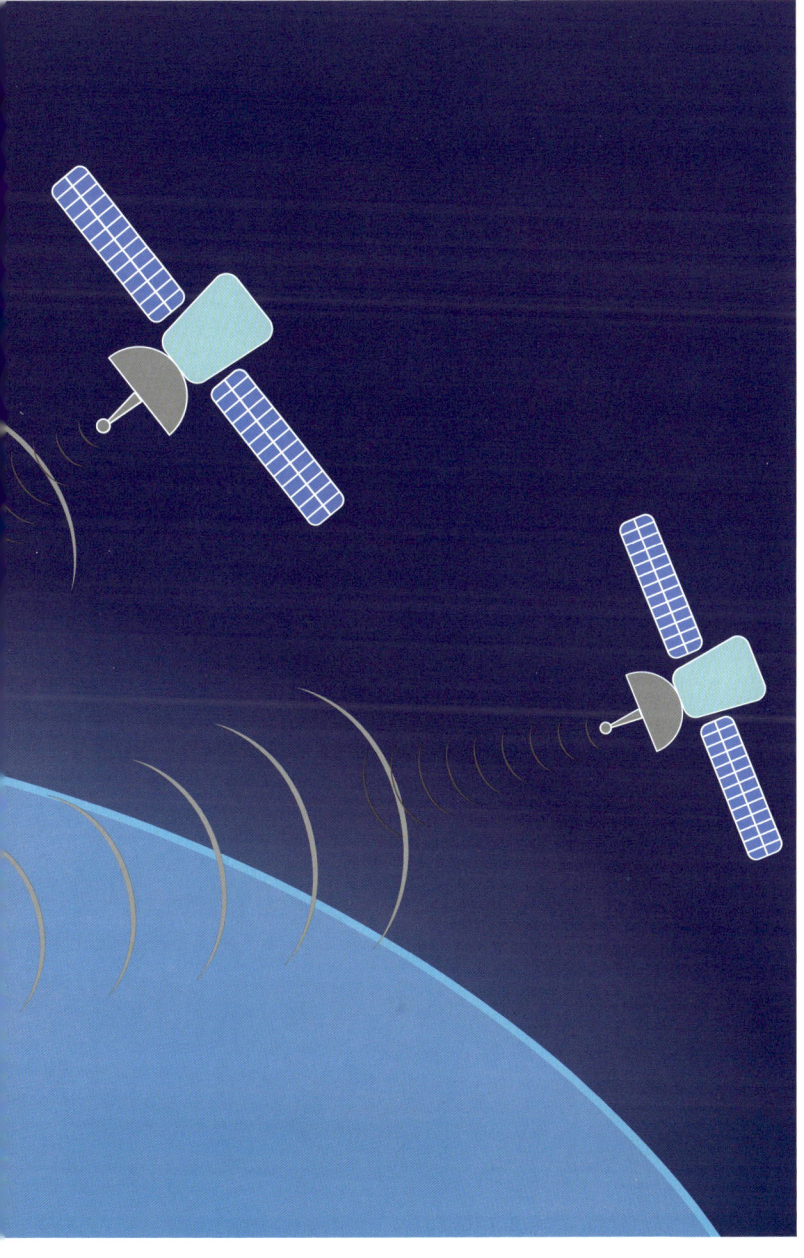

Do you want to know how far away the nearest gas station is? Do you want to know how far it is from school? Do you want to know where to find a fishing spot? You can find out using GPS!

Fiber Optics Uses Wave Signals to Transfer Information

Fiber optics is a process of transferring information via light waves using a special cable. Light travels in waves along bundles of thin, flexible glass fibers. The fibers are bundled inside the cable. If you have ever used cable TV, you have used fiber optics.

How are light waves used to send signals in a fiber optic cable? First, a transmitter changes information—such as a recording of a TV show—to digital code. The code is then transformed from numbers to flashes of light. The light waves enter one end of the cable. The cable is special. It allows the light to move freely through it without disrupting the wave pattern.

The other end of the cable is attached to a receiver. The receiver receives the light waves and translates them back to an electronic digital signal. Finally, the electronic signal is translated back to information the device can understand. You can then watch your favorite TV show!

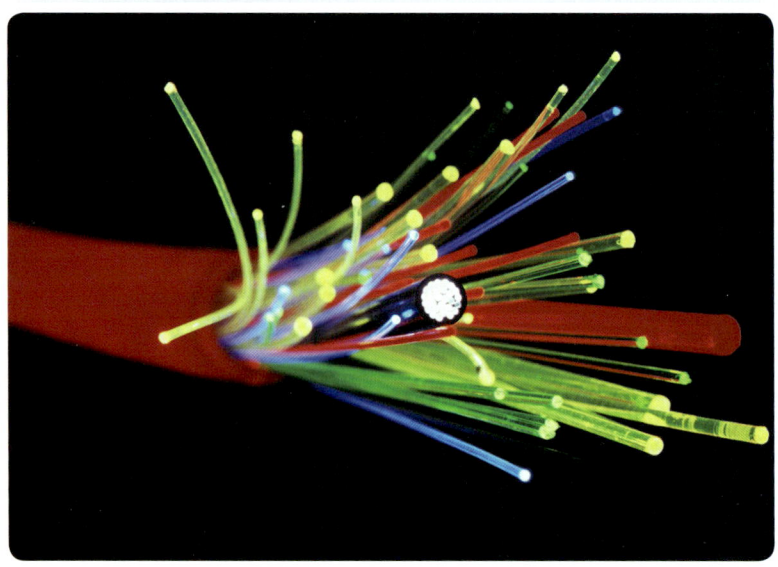

Fibers inside a cable carry light waves, which can then be transformed to information your device can understand.

Glossary

A

absorb, v. to take something in and contain it (18)

analog, adj. describing signals sent as continuous waves (34)

C

code, n. a pattern of symbols that can be used to communicate a message (29)

color, n. an aspect of light that enables otherwise identical objects to be distinguished from each other through the sense of sight (vision) (18)

crest, n. the highest part of a wave (3)

D

digital, adj. describing signals sent as separate bits and not continuous (35)

F

frequency, n. the number of times a wave peaks over a period of time (7)

I

information, n. types of data such as pictures, sounds, and messages (33)

intensity, n. the measurable strength or power of a vibration (7)

L

light, n. a form of energy that can transfer through empty space and can make things visible (13)

light source, n. an object that gives off its own light (14)

light wave, n. an energy disturbance that transfers, or radiates, light (15)

P

pattern, n. a regular or repeated way in which something occurs (29)

pitch, n. the quality of sound that is described as high or low and is related to a wave's frequency (7)

R

reflect, v. to bounce off of (16)

S

signal, n. a symbol sent over a distance (30)

sound wave, n. a transfer of energy through a material as it is disturbed by vibrations (5)

symbol, n. a visual object or mark that stands for something else (28)

T

transfer, v. to move from one place to another (33)

transparency, n. a property of matter that allows light to pass through it (16)

trough /trof/, n. the lowest part of a wave (3)

V

vibrate, v. to move back and forth quickly (5)

vibration, n. the motion of an object or material that is vibrating (5)

volume, n. the way humans perceive loudness from the intensity of a sound wave (7)

W

wave, n. a disturbance that transfers energy through matter or through space (1)

wave height, n. the vertical distance from the top of the crest to the bottom of the trough of a wave (3)

wavelength, n. the distance from one crest to the next crest of a wave (3)

Series Editor-in-Chief
E.D. Hirsch Jr.

Editorial Directors
Daniel H. Franck and Richard B. Talbot

Subject Matter Expert

Martin Rosenberg, PhD
Teacher of Physics and Computer Science
SAR High School
Riverdale, New York

Illustrations and Photo Credits

4X5 Collection / SuperStock: 40
Andrei Stanescu / Alamy Stock Photo: 30b
Andriy Popov / Alamy Stock Photo: 25b
Aurora Photos / Aurora Photos / SuperStock: Cover D, 2
Canva Pty Ltd / Alamy Stock Photo: 9
Christian Bäck / Mauritius / SuperStock: 22c
Cultura Creative (RF) / Alamy Stock Photo: 24
Cultura Limited / Cultura Limited / SuperStock: 1a, 1b, 17b
Don White / SuperStock: 17a
Elisabeth Burrell / Alamy Stock Photo: 26b
Erik Isakson / Blend Images / SuperStock: 23a
Gale S. Hanratty / Alamy Stock Photo: 16
Hajes / Alamy Stock Photo: 31
Image Source Plus / Alamy Stock Photo: 20b
Ingram Publishing / SuperStock: 30a
Inti St Clair / Blend Images / SuperStock: 33
Javier Larrea / age fotostock / SuperStock: 25d
Juice Images / SuperStock: 25a
Maskot / SuperStock: 26a
Matthias Lenke / F1 ONLINE / SuperStock: Cover C, 22a
OJO Images Ltd / Alamy Stock Photo: 32
Papilio / Alamy Stock Photo: 12
Radius / SuperStock: 13
Richard Maschmeyer / age fotostock / SuperStock: 28
Science Photo Library / SuperStock: 14, 25c
Terry Livingstone / Alamy Stock Photo: 22b
Tetra Images / SuperStock: 27
Valeriy Novikov / Alamy Stock Photo: Cover A, 29b
Visions of America / SuperStock: 34
Westend61 / SuperStock: 20a